D0991682

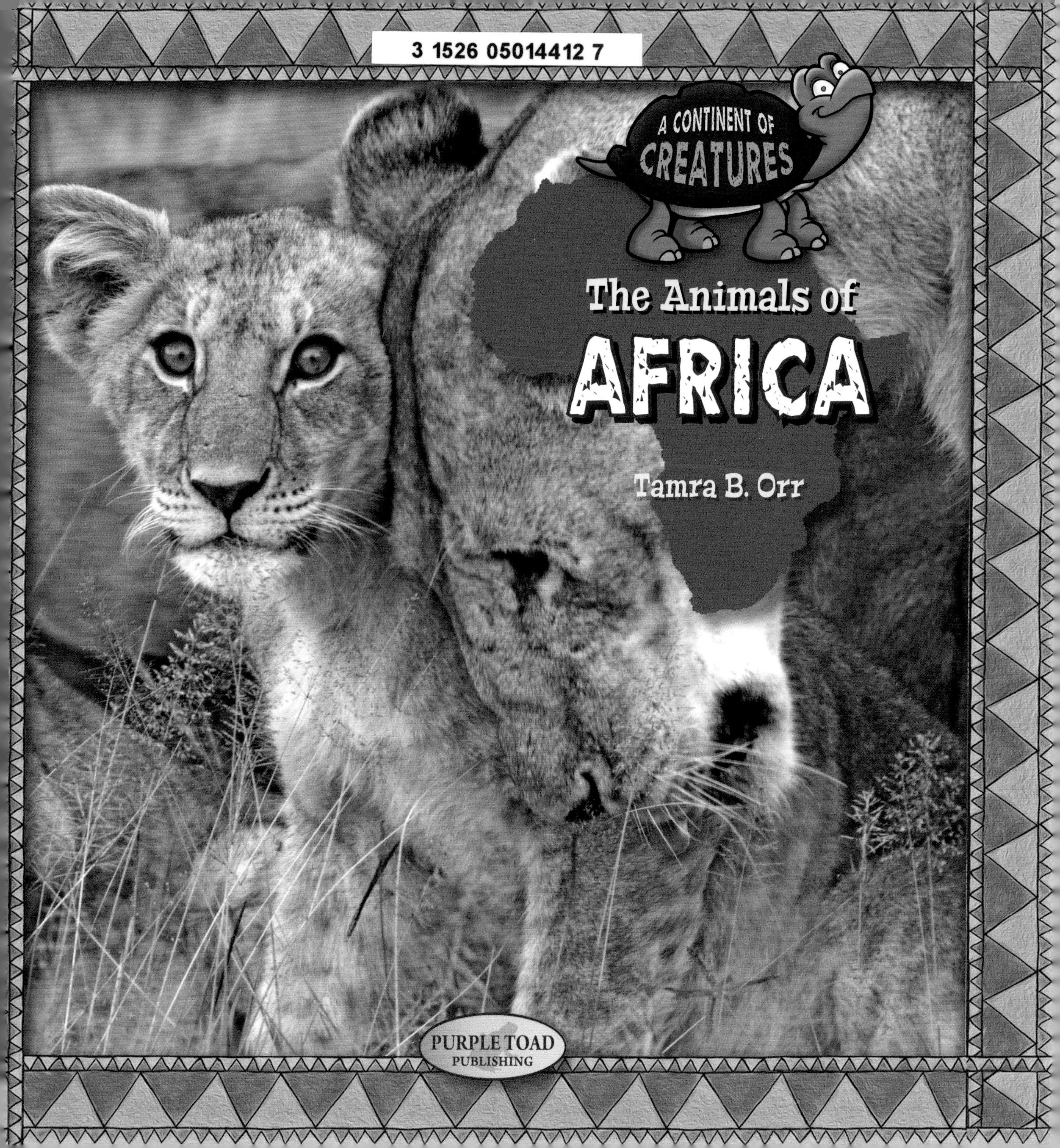

3 1526 05014412 7
A CONTINENT OF
CREATURES
The Animals of
AFRICA
Tamra B. Orr
PURPLE TOAD
PUBLISHING

Africa is a huge continent. It has 54 countries in it and covers 12 million square miles.

Welcome to Africa! Within its rain forests, deserts, and grasslands are animals of every kind imaginable. Let's get started meeting some of them!

Africa's grasslands are home to the zebra. Zebras are covered in black and white stripes to help hide them from the animals that hunt them. No two zebras have the same stripe pattern.

Africa has vast deserts like the Kalahari **(ka-la-HAH-ree)** and the Sahara **(sah-HAH-rah)**. This is where camels live. Their thick brown coat reflects sunlight, which helps them stay cool. Their humps hold fatty tissue so they can walk a long time without eating or drinking.

In the Moroccan (mor-AH-kin) heat, baby camels work hard to keep up with their moms.

Lions live in grasslands and plains. They stay in family groups called prides. These big cats love to sleep as much as 20 hours a day. Lions use their roar to talk to other lions, especially over long distances. A lion's roar can be heard up to five miles away.

The lion, however, is not the most dangerous animal in Africa. It is the hippopotamus **(hih-poh-PAH-tuh-mus)**. The hippo does not want anyone coming into its territory. It will charge with its very large teeth. Watch out!

The 4,000-pound hippo spends most of its time in the water. No wonder the name *hippopotamus* means "river horse."

The elephants of Africa are the world's largest living land mammals. They weigh up to seven tons. They send messages to each other over great distances by stomping. Their heavy feet make earth-shaking rumbles that other elephants can feel under their feet.

Elephants and giraffes live on the African savanna.

The giraffe is the world's tallest land animal (up to 14 feet tall). Giraffes can eat leaves from trees that other creatures can't reach. Their long tongues can carefully pull the leaves off the branches. They fill their bellies with up to 100 pounds of leaves in a day!

The cheetah's spots help it hide in the long African grasses.

Some of the world's fastest creatures are found here, but none are faster than the cheetah. The cheetah can run as fast as 70 miles per hour, helping it catch gazelles and small wildebeests.

Gazelles and wildebeests **(WIL-duh-beests)** share the same habitat as the cheetah. Gazelles are deer-like creatures. They are light and very quick. Wildebeests are large antelopes. They have hairy manes like a horse and curved horns.

Wildebeests and gazelles are always on the move, migrating from one place to another in search of fresh grass to eat.

Mandrills are the largest monkeys.

Chimpanzees are called "chimps" for short.

Chimpanzees **(chim-PAN-zees)** live in Africa's jungles and dry forests. These apes are very smart and use tools to do things like break open nuts. Mandrills are monkeys of the rain forest. They have pouches in their cheeks that they can stuff with food to eat later.

There are fewer than 1,000 mountain gorillas left in the world today.

Mountain gorillas may look fierce, but they are shy and like to stay in their forest homes. They love munching on plants, worms, and certain fruits. The biggest of the group becomes the leader. He is called a silverback for the gray hair he gets at about age 11 to 13.

Ostriches **(AH-strich-is)** are the world's largest birds. They cannot fly, but they can sprint at 43 miles per hour. A strong kick from their powerful legs may kill a lion.With their large eyes, they can see great distances.

When a gray parrot is frightened, it makes a growling sound.

The African gray parrot is not only beautiful but smart. It is able to imitate almost any sound, including the human voice. It can say whole sentences, and even chirp, beep, and meow.

Hiding along the banks of rivers are Nile Crocodiles (KROK-uh-dy-els). The largest reptile in Africa grows to over 16 feet long from nose to tail. They sink down under the water with only their eyes showing. They wait for an animal to come by for a drink. Then *bam!* They use sharp teeth and strong jaws to eat lunch.

Another sneaky creature is the black mamba snake. It is the most dangerous snake in Africa. It can bite up to 12 times a minute, and each bite has enough venom to kill a large animal.

Driver ants attack and eat termites.

African honeybees are also called killer bees.

Many fearsome insects dwell in Africa. African ants, or driver ants, are small but scary! Their pincers are so strong that some African tribes use their snapping heads like stitches to close up open wounds.

African honeybees can swarm and sting if animals, people, or even cars come too close to where they live.

The fattail scorpion lives in warm, dry areas.

African cichlids live in one of the deepest lakes in the world.

Eight-inch-long scorpions **(SKOR-pee-uns)** are found in the desert regions of North Africa. Venom from the black fattail scorpion is among the deadliest in the world.

The African cichlid **(SIH-klid)** is a fish of sparkling blue, yellow, and pink. You might even see one in your local aquarium!

Africa's rhinoceroses (ry-NAH-suh-rus-es) weigh up to 4,000 pounds. They stand 6 feet tall and are more than 12 feet long. Their mighty horns are made of the same stuff as our hair and fingernails.

Many amazing animals call Africa home. It's easy to see why. Africa has mountains, rain forests, savannas, rivers, and deserts — all on one continent!

FURTHER READING

Books

Berkes, Marianne. *Over in the Grasslands: On an African Savanna*. Nevada City, CA: Dawn Publications, 2016.

Owings, Lisa. *Meet a Baby Hippo*. Minneapolis, MN: Lerner Publications, 2015.

Rustad, Martha. *African Animals*. North Mankato, MN: Capstone Books, 2014.

Schaefer, Lola. *Run for Your Life! Predators and Prey on the African Savanna*. New York: Holiday House, 2016.

Web Sites

A to Z Kids Stuff: Africa

http://www.atozkidsstuff.com/africa.html

Africam: The Naledi Cat-EYE

http://www.africam.com/wildlife/naledi_cat_eye_live_wildlife_channel

Explore: African Watering Hole Animal Camera

http://explore.org/live-cams/player/african-watering-hole-animal-camera

National Geographic Kids: South Africa

http://kids.nationalgeographic.com/explore/countries/south-africa/#south-africa-johannesburg.jpg

GLOSSARY

aquarium (ah-KWAR-ee-uhm)—A glass container for fish.

cichlid (SIH-klid)—A type of freshwater fish with spiny fins.

continent (KON-teh-nent)—One of the great divisions of land.

desert (DEZ-ert)—A dry area.

grassland—Land covered in grasses and low green plants.

imitate (IM-eh-tayt)—Copy.

mammal (MAM-al)—Warm-blooded animal.

mane—Hair on an animal's neck.

plain—Flat area of land.

pouches—Pockets.

rain forest—Area where it rains daily.

savanna (suh-VAN-uh)—A flat grassland in a tropical area.

territory (TAIR-ih-tor-ee)—Area an animal lives in.

venom (VEN-um)—Poison produced by an animal.

PHOTO CREDITS: p. 1—David Dennis; p. 2—Fibercool; p. 8, 14—Bernard DuPont; p. 9—Sharmzpad; p. 10—Vince Smith; p.11—Joseph Eccheverria; p. 12—Afrika Force, Loren Javier; p. 16—Steve Slater; p. 17—Bill Love; p. 19—Marcel Sigg; p. 20—Frank Pecchino. All other photos—Public Domain. Every measure has been taken to find all copyright holders of material used in this book. In the event any mistakes or omissions have happened within, attempts to correct them will be made in future editions of the book.

INDEX

Printing 1 2 3 4 5 6 7 8 9

The Animals of Africa
The Animals of Antarctica
The Animals of Asia
The Animals of Australia
The Animals of Europe
The Animals of North America
The Animals of South America

ABOUT THE AUTHOR: **Tamra Orr is the author of hundreds of books for readers of all ages. She loves the chance to learn about faraway lands and to find out what it is like to live there—all from the comfort of her work desk. Orr is a graduate of Ball State University, and is the mother of four. She lives in the Pacific Northwest and goes camping whenever she gets the chance.**

Publisher's Cataloging-in-Publication Data
Orr, Tamra.
Africa / written by Tamra Orr.
p. cm.
Includes bibliographic references, glossary, and index.
ISBN 9781624692741
1. Animals—Africa—Juvenile literature. I. Series: A continent of creatures.
QL336 2017
591.96

Library of Congress Control Number: 2016937181

eBook ISBN: 9781624692758